ELEVATOR KNOWLEDGE

Sequence of Operation & Control Logic

By Teboho Ketso Kotsi

ISBN 978-1-9195281-0-6

First Edition

Printed in Ireland

Illustrations, block diagrams, and technical sketches by: Lesego Kendel

Publisher: **Teboho Ketso Kotsi**

DISCLAIMER:

The wiring diagrams, block diagrams, and technical illustrations contained in this book are provided for conceptual and educational purposes only.

They are simplified representations intended to explain operational logic, supervisory relationships, and system sequencing.

They do not represent the exact wiring configuration of any specific manufacturer or installation.

For installation, maintenance, commissioning, or fault-finding activities, the official wiring diagrams and technical documentation provided by the original equipment manufacturer must always be followed.

All work on elevator systems should be performed by properly trained and authorised personnel in accordance with applicable regulations, standards, and safety requirements.

The author and contributors shall not be held liable for any loss, injury, or damage resulting from reliance on the illustrative content of this publication.

Table of Contents

Dedications

This work is dedicated to the engineers, technicians, and apprentices who dedicate their lives to maintaining safe and reliable vertical transportation systems.

Behind every elevator journey are professionals whose knowledge, discipline, and commitment ensure that passengers travel safely between floors every day.

May this knowledge support those who continue advancing the elevator industry and preserving the standards of safety and professionalism that define this field.

About the Author

Teboho Kotsi is an elevator industry professional with over 22 years of hands-on experience in commissioning, maintenance, modernization, and technical support. His career has been defined by a passion for solving complex technical problems and a commitment to sharing knowledge. Based in Dublin, Ireland, Teboho writes to bridge the gap between highly technical concepts and practical, real-world applications, empowering technicians and engineers to excel in their fields.

His technical expertise is built on a strong educational foundation, including an N3 qualification in Electrical Engineering with distinctions in Engineering Science, an N4 National Certificate in Marketing Management, and a Lift Inspector certification earned Cum Laude. This blend of practical field experience and academic rigor informs every page of The Elevator Knowledge Series, a structured collection of volumes designed to build competence and confidence in professionals across the industry.

Teboho approaches writing with the same methodical precision he uses for fault diagnosis, turning complex systems into clear, accessible knowledge. He lives by the mottos, "Love what you do and do what you love," and "We rise by lifting others"—principles that guide his work as an author, mentor, and engineer. When he's not on-site or writing, he enjoys mentoring the next generation of technicians and staying connected with his family in South Africa, including his beloved cat, Socks.

You can connect with Teboho on LinkedIn at http://www.linkedin.com/in/teboho-ketso-kotsi-08986964

PREFACE

Elevator systems are often approached as collections of components: doors, drives, motors, controllers, and safety circuits. Engineers are trained to understand parts, specifications, and procedures. What is less often taught is how these parts think together as a system.

This book was written to address that gap.

The purpose of this volume is not to repeat manufacturer documentation or regulatory standards. It is to clarify the logic governing elevator behaviour from the moment a call is registered until the doors open at the destination floor.

Elevators do not move because a button was pressed. They move because a sequence of conditions has been satisfied. Understanding this sequence changes how Engineers approach troubleshooting, commissioning, and supervision. It shifts the mindset from component replacement to logical diagnosis.

Throughout this book, motion is treated as the outcome of verified conditions. This principle is not philosophical language; it reflects how elevator systems are designed. Movement occurs only when integrity has been established.

This text is based on field experience. It is written for Engineers who work with live systems and real consequences. Safety is not discussed as an abstract requirement but as a continuous condition embedded within every stage of operation.

If this book strengthens clarity of reasoning and improves diagnostic discipline even slightly, it will have served its purpose.

HOW TO USE THIS BOOK

This book follows the actual operational sequence of an elevator system. Each chapter represents a stage in the logic flow from call registration to door opening and special operating modes.

It may be read in two ways.

First, it may be read sequentially, beginning with terminology and call registration, progressing through door operation, safety supervision, drive enable, motion, levelling, and concluding with special modes such as fire service operation. This approach builds a structured understanding of how decisions flow through the system.

Second, it may be used as a reference. Engineers encountering a specific issue may turn directly to the relevant chapter, such as door operation, safety circuits, or brake control, and review the logical conditions required at that stage.

When using the book in the field, the key question to apply consistently is:

Which condition required for this stage has not been satisfied?

This question supports disciplined troubleshooting and reduces unnecessary component replacement.

This volume assumes a basic familiarity with elevator components and terminology. It is written primarily for Engineers working in installation, maintenance, commissioning, inspection, or technical supervision roles.

It is not intended to replace standards, statutory requirements, or manufacturer instructions. Those documents remain authoritative. This book supports an understanding of how those requirements integrate into operational logic.

FOREWORD

Elevator systems are among the most quietly trusted safety-critical systems in modern buildings. Millions of journeys occur daily without incident, and this reliability is rarely noticed.

That reliability does not occur by chance. It depends on disciplined engineering, structured supervision, and clear decision-making under pressure.

This book presents elevator operation as a sequence of controlled decisions rather than a collection of independent mechanisms. It emphasises that motion is subject to verified conditions and that safety supervision is continuous rather than optional.

The value of this approach lies in its practicality. When Engineers understand the order in which conditions must be satisfied, fault diagnosis becomes logical rather than reactive. Symptoms are interpreted within context, and corrective actions become deliberate rather than speculative.

This work is particularly relevant for Engineers who recognise that technical competence alone is not sufficient. Consistency of judgement, clarity of reasoning, and respect for safety hierarchy are equally important.

The chapters that follow trace the operational flow of a lift system step by step. By doing so, they encourage Engineers to think in sequences, not fragments.

ACKNOWLEDGEMENTS

This book is the result of years of field experience, observation, and disciplined engineering practice.

I extend my appreciation to the Engineers, mentors, colleagues, and technicians whose work ethic and professional standards have influenced my own approach to system integrity and structured reasoning.

Special recognition is due to those who operate within safety-critical environments daily, often under pressure, yet remain committed to verification and responsible decision-making.

I would also like to acknowledge **Lesego Kendel**, who drafted all illustrations, block diagrams, and technical sketches included in this volume. His precision and visual clarity have significantly strengthened the structured presentation of the operational framework. Lesego is a professional photographer, and his disciplined approach to visual composition translated seamlessly into technical diagram development.

Engineering knowledge is not created in isolation. It is refined through experience, reflection, and collaboration.

Teboho Ketso Kotsi

CHAPTER 1
Elevator Operation as a System of Verified Conditions

1.1 Elevators as Decision Systems

An elevator is often described by its components: controller, motor, drive, doors, safety devices, and landing equipment. While understanding these parts is essential, the system cannot be fully understood by examining components in isolation.

An elevator is a sequence of controlled decisions.

From the moment a call is registered until doors open on the destination floor, the system evaluates conditions in a defined order. Each stage depends on the integrity of the stage before it. The system advances only when specific requirements are satisfied.

This ordered progression enables elevators to operate reliably in public environments without continuous human supervision.

1.2 Movement as a Consequence, Not a Trigger

It is common for passengers to assume that pressing a button initiates movement. In reality, the button press begins a chain of evaluations.

Before propulsion occurs, the system must confirm:

- Valid call registration
- Established direction
- Closed and locked doors
- Complete safety circuit
- Drive enable conditions
- Brake release confirmation

Motion is the consequence of verified supervisory continuity across multiple control layers.

This distinction fundamentally alters diagnostic interpretation. When motion does not occur, the correct question is not what has failed, but which required condition remains unsatisfied.

1.3 Permission-Based Logic

Elevator systems operate on what may be described as permission-based logic.

At each stage, the system asks whether it is permitted to proceed. If any required input is absent or invalid, advancement is denied. This denial is not an error; it is a protective response.

Permission-based logic ensures that propulsion cannot override supervision. It prevents movement when safety conditions are incomplete. It reinforces the hierarchy in which supervision governs action.

This framework provides Engineers with a practical mental model. Instead of chasing symptoms, they can evaluate which permission has not been granted.

1.4 Continuous Supervision

Safety in elevator systems is not an isolated function. It is continuous.

The safety circuit does not activate at the moment of danger; it remains active throughout operation. Door interlocks, overspeed protection, limit devices, and supervisory inputs are monitored before and during movement.

Continuous supervision means that protection is embedded in the system's architecture. The elevator does not rely solely on a final emergency response; it prevents unsafe states from occurring.

Engineers must therefore view safety not as a feature, but as a structural condition that governs all motion.

1.5 Sequence Thinking

When diagnosing faults, sequence thinking becomes essential.

If the system fails to move, the cause will lie in one of the earlier verification stages. If doors reopen unexpectedly, the explanation lies in supervisory logic responding to a missing or interrupted input.

Sequence thinking replaces guesswork with order.

Instead of replacing components based on probability, Engineers can evaluate:

- Which stage should be active?
- Which input is required at this stage?
- Has that input been confirmed?

This approach reduces unnecessary part replacement and improves diagnostic accuracy.

1.6 Professional Responsibility

Elevator systems operate in environments where the public assumes safety without understanding the mechanisms behind them. Engineers are responsible for maintaining that trust.

Clarity of reasoning is therefore not optional. Misinterpreting supervision logic or bypassing protective conditions compromises both safety and professional integrity.

This book does not attempt to replace standards or manufacturer documentation. Instead, it clarifies the logical structure that underpins them. When Engineers understand the order of conditions and the hierarchy of supervision, they are better equipped to make disciplined decisions under pressure.

1.7 Transition to Operational Stages

With the foundation of permission-based logic established, the operational sequence now begins formally with call registration and direction determination. The next chapter examines how the system evaluates input requests and prepares for progression through the sequence.

FIGURE 1.1 – Permission-Based Operation Model

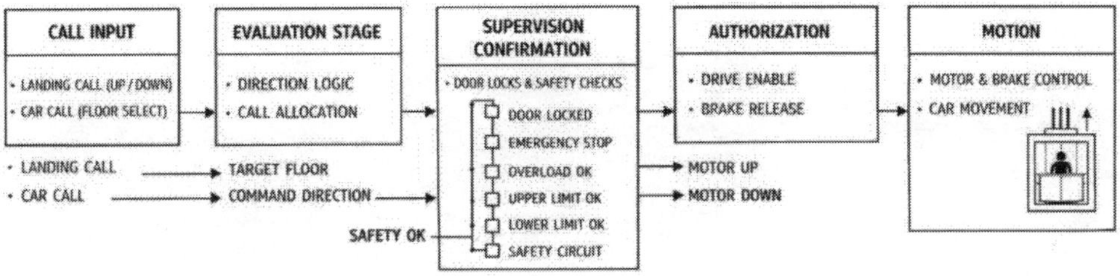

Figure 1.1 – Sequential Motion Permission Model.

This figure illustrates that elevator motion results from sequentially verified conditions, where each stage must grant permission before progression can occur.

CHAPTER 2
Call Registration & Direction Determination

FIGURE 2.1 – Integrated Sequence of Operation

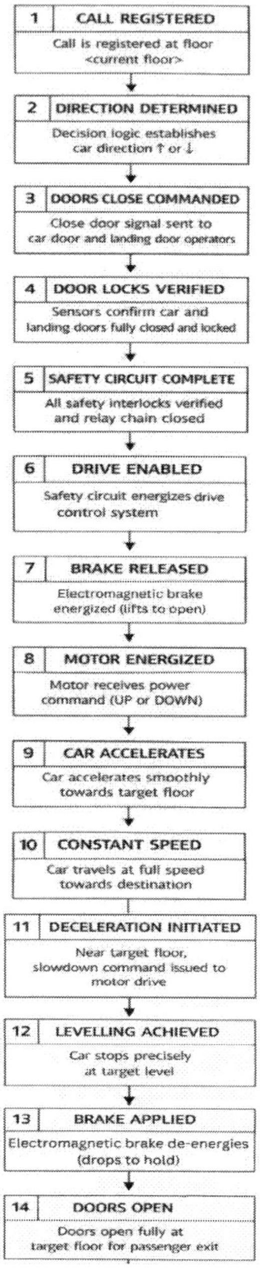

Figure 2.1 - Integrated Operational Sequence.

This diagram represents the full operational flow from call registration through propulsion levelling and door opening, with each stage dependent on the completion of the preceding condition.

2.1 The Beginning of the Operational Sequence

The operational sequence begins when a call is registered.

As shown in Figure 2.1, call registration represents the first formal input into the system's decision chain. At this stage, no movement is initiated. The system records a request and prepares to evaluate it.

Call registration may originate from:

- A landing call button
- A car operating panel
- A destination control terminal
- A supervisory command

Regardless of origin, the controller must first confirm that the call signal is valid. Electrical integrity, button condition, and communication pathways must all function correctly for the request to be acknowledged.

Invalid call registration prevents progression beyond the initial stage.

2.2 Call Validation

Not every signal results in action. Before a request is accepted into the dispatch logic, the controller verifies:

- That the lift is in a mode permitting normal operation
- That the requested floor exists within programmed limits
- That no overriding mode (inspection, fire service, or independent service) is active

Invalid or restricted inputs are ignored or overridden by higher-priority logic.

At this stage, the system remains stationary. It is evaluating eligibility, not authorizing motion.

2.3 Group Control and Allocation

In multi-lift installations, call registration triggers an allocation decision.

The group controller evaluates:

- Current car positions
- Direction of travel
- Load conditions
- Existing commitments

The objective is not simply to respond, but to respond efficiently.

In a simplex system, this stage is straightforward. In duplex or triplex systems, allocation logic becomes a dynamic optimization problem.

Once a car is assigned, the system advances to direction determination.

2.4 Determining Direction

Direction is not arbitrary. It is determined by:

- Current car position
- Registered calls
- Operational mode
- Traffic algorithm

The controller selects an initial direction of travel based on internal rules designed to minimize delay and avoid conflict.

Importantly, direction determination prepares the system for progression but does not authorise motion.

If conflicting conditions exist, such as simultaneous up and down calls on the same floor, the controller resolves priority based on programmed logic.

2.5 Mode Influence on Call Logic

Operating modes significantly alter call behavior.

Examples include:

- Independent Service: Landing calls ignored
- Fire Service Phase 1: Automatic recall overrides normal calls
- Inspection Mode: All passenger calls disabled
- Priority Mode: Certain inputs override others

Understanding the influence of mode is essential during diagnostics. Apparent call failures are often due to mode restrictions rather than hardware faults.

Engineers must confirm the operating mode before investigating the call circuitry.

2.6 Call Cancellation and Retention

Once registered, a call remains active until:

- The car services the requested floor
- The call is cancelled by logic
- The system is reset

Improper cancellation behavior may indicate:

- Faulty pushbutton contacts
- Stuck inputs
- Controller memory anomalies
- Wiring inconsistencies

Accurate interpretation requires understanding at which stage the sequence halted.

2.7 Diagnostic Focus

When investigating call-related issues, structured reasoning prevents misdiagnosis.

Engineers should ask:

- Was the call registered?
- Was it allocated to a car?
- Was direction established?
- Was the mode preventing the response?

If the system does not proceed beyond call registration, the fault lies within validation, allocation, or supervisory logic - not propulsion.

This distinction avoids unnecessary investigation of the drive or motor.

2.8 Transition to Door Operation

Once a car has been allocated and a direction determined, movement remains conditional. Before propulsion can be authorised, door closure and lock verification must be confirmed. The next chapter examines the mechanical and electrical requirements of door integrity within the operational sequence.

As shown in Figure 2.1, progression from Stage 2 to Stage 3 requires that doors be commanded to close and subsequently verified as locked before the sequence can advance.

CHAPTER 3
Door Operation and Closure Verification

FIGURE 3.1 - Door Closure Verification Flow

Figure 3.1 – Door Closure Verification Logic.
This figure illustrates the decision logic used to confirm mechanical closure and electrical lock continuity before safety circuit evaluation.

3.1 The Operational Requirement for Closed Doors

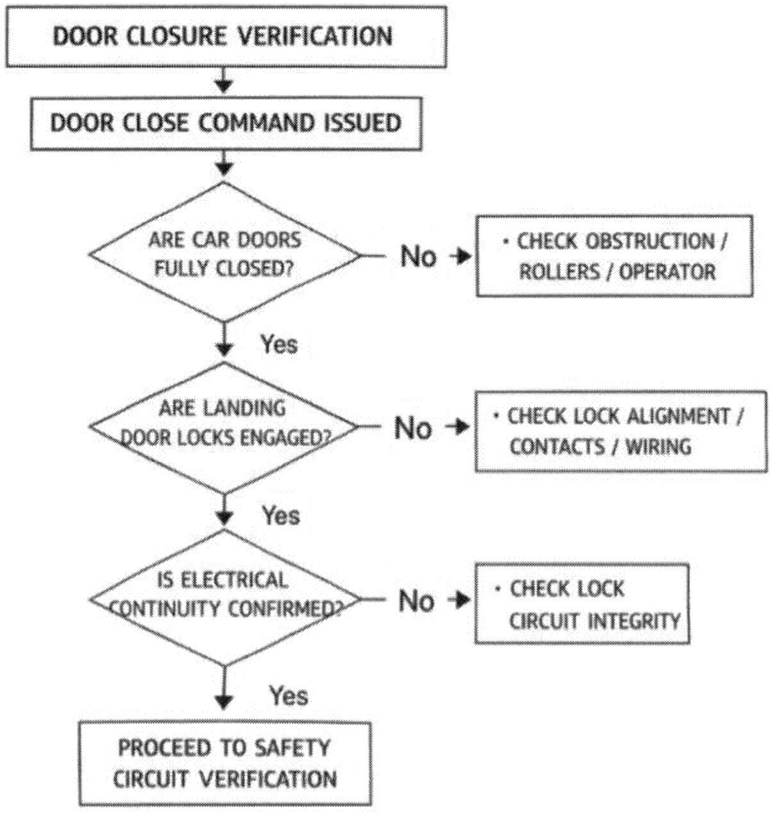

Following call allocation and direction determination, the system must secure the car before propulsion can be authorised.

As illustrated in Figure 2.1, the transition from direction establishment to propulsion passes through door closure and locking verification. This stage is not procedural formality; it is structural necessity.

The elevator controller issues a door-close command. However, issuing the command is not sufficient. The system must confirm that:

- Car doors are physically closed
- Landing doors are fully engaged
- Mechanical interlocks are secured
- Electrical lock contacts indicate continuity

Until these confirmations are present, the sequence cannot progress.

3.2 Mechanical and Electrical Interdependence

Door systems combine mechanical movement and electrical supervision.

Mechanical components include:

- Door panels
- Tracks and rollers
- Hanger assemblies
- Closing mechanisms

Electrical supervision includes:

- Door closed limit switches
- Lock contacts
- Door operator feedback signals

A door may appear physically closed while failing to produce a valid electrical confirmation. Conversely, electrical signals may fluctuate if mechanical alignment is marginal.

Engineers must recognise that both layers must align for progression.

3.3 Door Lock Verification

Door security is confirmed; it is never assumed.

Lock verification typically involves:

- Electrical continuity through the landing door lock contacts
- Confirmation that the car door lock contacts are engaged
- Monitoring for interruption during movement

Only when the entire lock chain indicates integrity does the system permit the next stage of evaluation.

Door lock failure is one of the most common causes of non-movement conditions.

3.4 Common Door-Related Failure Modes

When door verification fails, the system remains stationary.

Common causes include:

- Mechanical obstruction
- Misaligned landing lock
- Worn rollers
- Door operator timing faults
- Broken lock contact
- Intermittent wiring
- Contaminated switch contacts

A structured approach prevents unnecessary drive or controller intervention.

If lock continuity is absent, propulsion is intentionally withheld.

3.5 Intermittent Door Faults

Intermittent door faults present particular diagnostic challenges.

Symptoms may include:

- Lift operates normally, then fails without warning
- Door close cycle repeats
- Movement was denied after several successful trips

These conditions often indicate marginal lock alignment or fluctuating electrical continuity.

Engineers should avoid immediate component replacement. Instead:

- Observe lock engagement physically
- Monitor contact continuity under vibration
- Inspect mechanical wear

Intermittency is frequently mechanical rather than electronic.

3.6 Mode Influence on Door Logic

Certain modes modify door behaviour:

- Inspection mode may allow limited movement with bypassed landing lock circuits (depending on jurisdiction)

- Fire service operation changes door dwell and recall logic
- Independent service ignores landing calls

Understanding the active mode prevents misinterpretation of the door response.

3.7 Diagnostic Focus

When a lift refuses to move after call allocation:

1. Confirm the door-close command issued
2. Confirm mechanical closure
3. Confirm lock engagement
4. Confirm electrical continuity

Failed door verification halts progression prior to safety circuit evaluation.

The correct question remains:

- Which state has not been satisfied?

<u>Field Scenario – Case Study</u>

<u>Condition:</u>

- Lift registers calls normally. Direction is established. Doors appear fully closed. The car does not move.

<u>Initial Assumption (Common Error):</u>

- Drive fault or controller malfunction.

<u>Structured Evaluation:</u>

- Confirm door-close command issued.
- Observe whether the door operator completes the closing cycle.
- Inspect mechanical closure.
- Verify that door panels are fully aligned and not resting against an obstruction.
- Check landing door lock engagement.
- Confirm that the mechanical lock has fully engaged and is not marginally misaligned.
- Verify electrical continuity.
- Measure continuity across the lock circuit while gently applying pressure to the landing door.

Observation:

- Continuity drops intermittently when the door panel is lightly pressed.

Diagnosis:

- Marginal landing door lock alignment is causing an intermittent open circuit within the safety chain.

Corrective Action:

- Realign the landing lock mechanism and verify stable continuity under vibration.

3.8 Transition to Safety Circuit Verification

With door closure verified, the system expands supervision beyond access control. The next stage requires full safety circuit continuity before motion permission can be granted. The following chapter examines safety circuit architecture and supervisory hierarchy.

As shown in Figure 2.1, door verification precedes complete confirmation of the safety chain. The next stage expands supervision beyond doors to include overspeed devices, limit switches, inspection controls, and emergency stop inputs.

Safety circuit integrity now becomes the governing condition.

CHAPTER 4
Safety Circuit and Permission for Motion

4.1 From Door Verification to System-Wide Supervision

Door locks represent only one segment of the supervision structure.

Once door closure is confirmed, the controller evaluates the entire safety circuit. This circuit functions as a continuous chain of normally closed protective devices. Any interruption prevents propulsion.

As shown in Figure 2.1, progression beyond door verification depends on the complete safety chain being in place.

This stage expands supervision beyond doors to include:

- Overspeed governor contacts
- Final limit switches
- Inspection switches
- Emergency stop inputs
- Buffer switches

Additional protective devices, depending on the configuration.

If any element is open, movement is denied.

4.2 Structure of the Safety Circuit

A safety circuit, often referred to as the safety chain, is a series-connected supervisory loop. Its integrity depends on uninterrupted continuity across all protective devices.

Each device contributes one condition to the chain. The controller monitors the entire loop rather than individual devices in isolation.

This design ensures that:

- A single open contact halts progression
- Propulsion cannot occur if protection is compromised
- Supervision remains active throughout the operation

FIGURE 4.1 – Safety Circuit Supervision Architecture

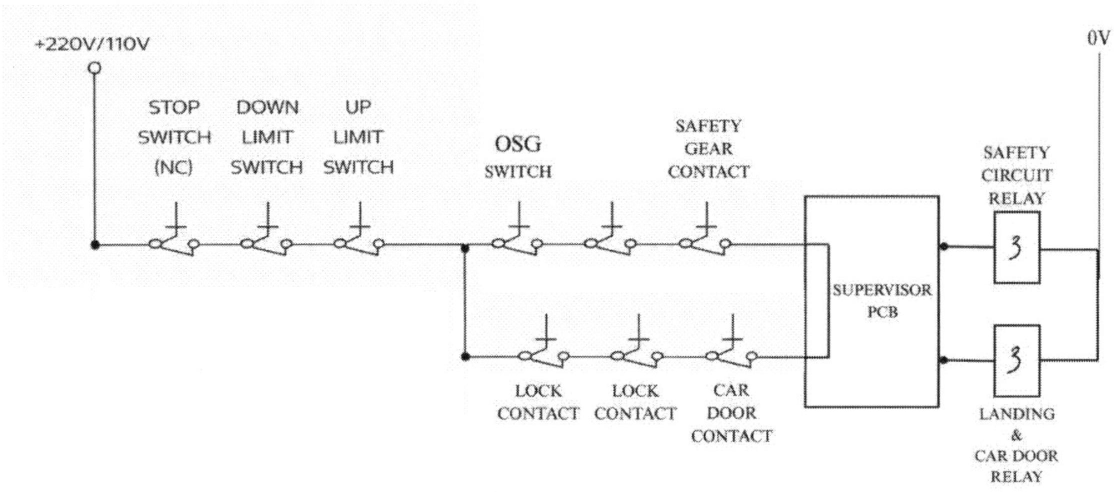

Figure 4.1 - Simplified representation of the safety supervision chain.

This schematic shows the series-connected safety circuit, in which continuity across all protective devices must be established before motion permission is granted.

Note: This representation is conceptual. The official OEM wiring diagram must be followed for installation and fault-finding.

4.3 Continuous Monitoring

The safety circuit is not verified once and forgotten. It is monitored continuously.

If continuity is lost during motion:

- The safety relay drops
- Drive power is removed
- Brake is applied
- The car decelerates to a stop

This behaviour reflects the supervisory hierarchy. Propulsion always remains subordinate to safety supervision.

4.4 Common Safety Circuit Failures

When safety continuity is absent, the system will not advance to drive enable.

Common causes include:

- Misaligned door lock contact
- Tripped overspeed device
- Activated inspection switch
- Open emergency stop
- Broken conductor
- Loose terminal connection

Engineers must identify which device has opened the chain rather than assuming controller failure.

4.5 Intermittent Safety Chain Interruptions

Intermittent interruptions may present as:

- Lift moves briefly, then stops
- Movement denied after vibration
- Random drop-outs

These faults often indicate:

- Loose wiring
- Oxidised contacts
- Marginal switch alignment
- Mechanical shock affecting contact integrity

Observation under operating conditions may reveal more than static testing.

Field Scenario – Case Study

Condition:

- Lift intermittently stopping between floors with passengers inside.

History:

- The variable-frequency drive was previously replaced due to repeated entrapments. Lift was subsequently taken out of service due to ongoing reliability concerns.

Observed Behaviour:

- The lift would travel normally.
- During motion, propulsion would suddenly cease.
- The main drive contactors dropped out while the car was in motion.
- The error log consistently indicated a drive fault.

Because the fault log pointed to the drive, it had already been replaced before further investigation.

Structured Investigation

Initial verification confirmed:

- Visible drive fault codes at rest.
- Safety circuit did not indicate an open condition.
- No recorded door lock faults.

However, during operation, observation revealed that the main drive contactors were de-energising mid-travel.

This behaviour suggested that the control voltage supply to the contactor coil was being interrupted.

Technical Findings:

Voltage measurement across the safety circuit revealed elevated resistance.

Further inspection identified high resistance across door lock contacts within the safety chain.

The circuit was not fully open. It was resistive. This distinction was critical. Because the circuit was not open, the controller did not register a lock fault. However, the increased resistance reduced coil voltage to the main drive contactors sufficiently to cause them to drop out during motion.
The drive was not at fault. It was losing its control voltage supply.

Corrective Actions:

- Cleaned and restored door lock contacts
- Verified stable continuity under vibration
- Confirmed full voltage present at the main drive contactor coil
- Conducted repeated travel tests under load

Following corrective action, contactors remained energised during motion. No further entrapments occurred.

Engineering Principle Reinforced

A high-resistance safety circuit fault can mimic drive failure.

Error logs reflect symptoms, not root cause.

Safety circuit evaluation must include voltage integrity under load, not continuity alone.

Structured reasoning prevented further unnecessary component replacement.

4.6. Diagnostic Sequence for Safety Interruptions

When safety continuity is lost:

- Confirm safety relay status
- Identify which segment of the chain is open
- Inspect mechanical alignment
- Verify electrical continuity under load
- Check recent maintenance or adjustment work

Systematic isolation prevents unnecessary part replacement.

4.7 Safety Circuit and Drive Enable

Completion of the safety chain does not immediately cause movement. It permits the controller to consider propulsion.

At this stage:

- Drive enable signals may be issued
- Brake control logic may be prepared for release
- Propulsion remains conditional

Safety continuity grants eligibility, not automatic action.

4.8 Transition to Drive and Brake Control

After safety circuit integrity is confirmed, the system becomes eligible for propulsion. The next chapter examines drive enable logic, brake control coordination, and the structured initiation of motion.

As shown in Figure 2.1, the next stages involve:

- Drive enable
- Brake release
- Motor energisation

The relationship between these elements determines controlled motion.

CHAPTER 5
Drive Enable, Brake Control, and Motion Initiation

5.1 From Supervision to Propulsion

With safety circuit continuity verified, the system transitions from supervision to propulsion preparation.

At this stage, the controller has confirmed:

- Valid call allocation
- Established direction
- Closed and locked doors
- Complete safety chain continuity

Only now does the controller issue propulsion-related signals.

Importantly, propulsion occurs in a defined order. The drive, brake, and motor do not act simultaneously or independently. Their interaction is coordinated.

5.2 Drive Enable

The drive enable signal is the controller's authorization for the inverter or motor control system to prepare for torque production.

The drive enable signal typically confirms:

- Safety relay energised
- Direction command valid
- No fault conditions active
- Internal drive diagnostics satisfied

When the drive enable signal is absent, propulsion cannot begin even if the safety circuit is complete.

The drive enable signal authorises torque production but does not initiate motion independently.

FIGURE 5.1 – Propulsion Control Relationship

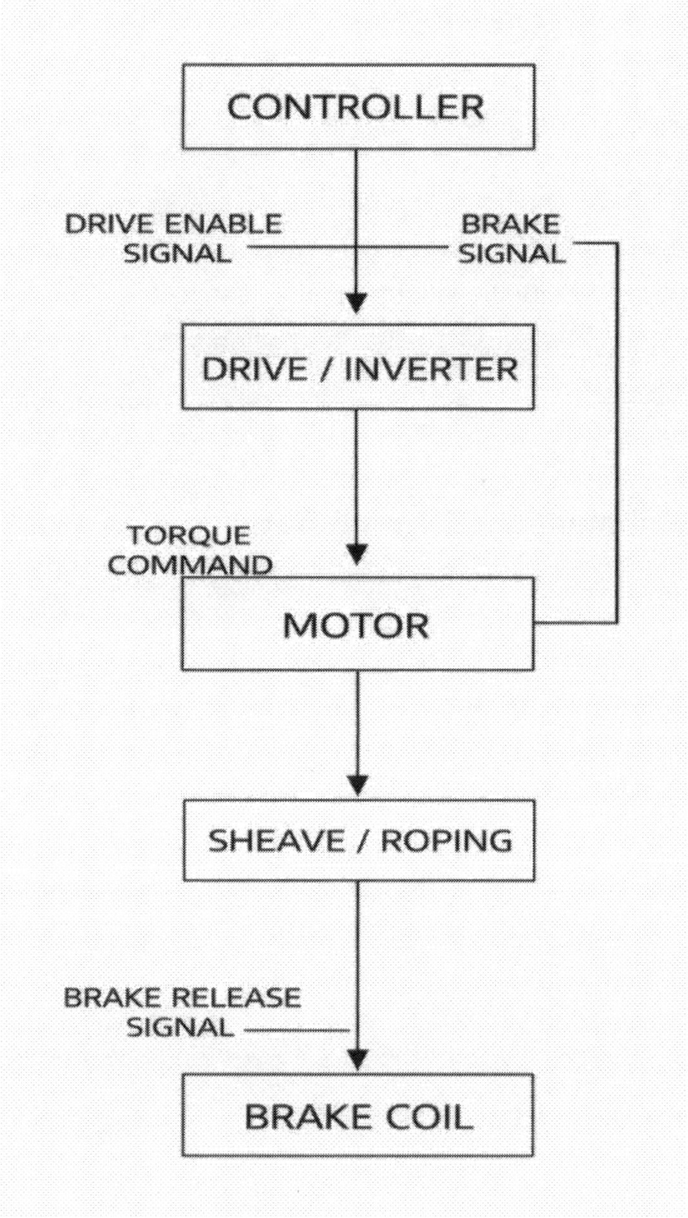

Figure 5.1 - Relationship between controller, drive, brake, and motor during propulsion initiation.

This diagram shows the interactions among the controller, drive enable signal, brake release, and motor torque during motion initiation.

5.3 Brake Release

The brake must release before rotation occurs.

The brake is typically spring-applied and electrically released. When the controller confirms drive readiness, it energises the brake coil.

Proper sequencing ensures:

- The brake releases only when torque is available
- The brake applies immediately if supervision is lost
- Mechanical holding integrity is maintained when stationary

A brake that fails to release prevents movement even when the drive enable signal is present.

5.4 Torque Establishment and Motor Rotation

Once the drive enable signal and brake release are confirmed, the drive produces controlled torque.

Modern variable frequency drives manage:

- Acceleration profile
- Torque ramp-up
- Current limitation
- Slip compensation

Acceleration must remain smooth and controlled to prevent shock loading.

The motor does not reach full speed immediately. Motion begins gradually.

5.5 Common Propulsion Fault Scenarios

When propulsion fails to occur despite completed supervision, possible causes include:

- Drive enable signal absent
- Drive internal fault
- Brake coil failure
- Insufficient brake voltage
- Mechanical brake seizure
- Motor winding issue

Engineers must determine which stage has not progressed.

If the drive enable signal is present, but the brake does not release, motion is mechanically blocked.

If the brake releases but the torque is absent, the drive may not be producing output.

5.6 Intermittent Motion Issues

Intermittent motion problems may present as:

- Jerking start
- Delayed movement
- Sudden drop after brief motion
- Random propulsion refusal

These may indicate:

- Marginal brake coil supply
- Loose motor terminals
- Drive parameter instability
- Overcurrent detection

Observation of sequencing under test conditions often reveals the failing stage.

5.7 Hierarchy During Motion

Once motion begins, supervision does not cease.

The safety circuit continues monitoring. If continuity is interrupted:

- Drive output is cut
- Brake is applied
- Car decelerates

This supervisory hierarchy remains intact throughout propulsion.

5.8 Transition to Controlled Travel

With propulsion initiated, controlled travel must follow a regulated speed profile. The next chapter examines acceleration, constant-speed operation, and deceleration phases in motion control.

As illustrated in Figure 2.1, the next stages involve:

- Acceleration profile
- Constant speed travel

- Deceleration initiation

Understanding controlled travel requires examining speed regulation and levelling precision.

Field Scenario - Case study (Drive Enable Signal Present, No Motion)

Condition:

Lift responds to calls. Doors close. Safety circuit verified. Drive enable signal present. Brake coil energised. Motor produces an audible hum. No movement occurs.

Initial Assumption (Common Error):

Drive malfunction or inverter failure.

Structured Evaluation:

1. Confirm that the drive enable signal is present at the drive terminals.
2. Confirm brake coil voltage is stable and within specification.
3. Observe brake release mechanically.
4. Measure the torque output current during the command.

Observation:

Brake coil energised, but the mechanical brake arm not fully releasing. Motor torque insufficient to overcome the partial mechanical hold.

Diagnosis:

Mechanical brake seizure due to contamination and insufficient clearance.

Drive and control signals functioning correctly.

Corrective Action:

- Clean brake assembly
- Adjust mechanical clearance
- Verify full brake release travel
- Confirm smooth motion under load

Engineering Principle Reinforced

Propulsion failure is not always electrical.

The drive enable signal and the brake release must be verified mechanically and electrically.

CHAPTER 6
Speed Control, Acceleration, and Deceleration Profiles

6.1 From Torque to Controlled Travel

Once the torque is established and the brake has released, the car begins to move.

However, motion is not binary. Elevators do not transition instantly from stationary to full speed. Travel occurs in controlled phases governed by programmed speed profiles.

These phases include:

- Acceleration
- Constant speed travel
- Deceleration
- Levelling approach

Each phase is precisely managed by the drive and controller.

6.2 Acceleration Phase

During acceleration, the drive gradually increases the motor frequency. Torque is applied in a controlled ramp to avoid:

- Mechanical shock
- Passenger discomfort
- Rope slip
- Overcurrent conditions

Acceleration rate is determined by:

- Drive parameters
- Rated speed
- Load conditions
- System tuning

Acceleration must be smooth and consistent. Abrupt transitions often indicate drive tuning errors or brake timing issues.

FIGURE 6.1 – Elevator Speed Profile Curve

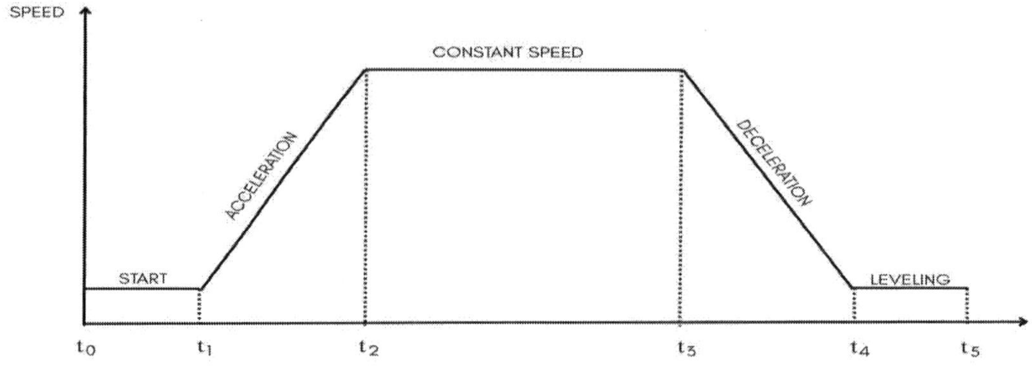

Figure 6.1 – Elevator Speed Profile Curve.

This figure illustrates the controlled speed phases of elevator travel, including acceleration, constant velocity, deceleration, and levelling.

6.3 Constant Speed Phase

After acceleration, the car enters constant speed travel.

In this phase:

- Motor frequency stabilizes
- Torque output matches load demand
- System monitors supervisory inputs continuously

Constant speed travel is typically uneventful when systems are properly tuned.

However, irregularities during this phase may indicate:

- Drive instability
- Load imbalance
- Encoder feedback errors
- Mechanical vibration

A stable, constant speed indicates a proper parameter configuration.

6.4 Deceleration Phase

Deceleration begins before reaching the destination floor.

The controller calculates when to initiate a slow-down based on:

- Position feedback
- Travel speed
- Load
- Programmed deceleration curve

The drive gradually reduces frequency, allowing the car to approach the landing at reduced velocity.

Incorrect deceleration timing may cause:

- Overtravel
- Undershoot
- Harsh stopping

Proper deceleration ensures a smooth transition into levelling.

6.5 Levelling Phase

Levelling is the final low-speed adjustment before the brake is applied.

At this stage:

- The car approaches creep speed
- Position sensors confirm floor alignment
- Minor adjustments are made

Precision levelling depends on:

- Encoder accuracy
- Drive tuning
- Mechanical condition
- Rope stretch compensation

Inaccurate levelling is rarely a simple motor issue. It often reflects feedback calibration or mechanical wear.

6.6 Feedback Devices and Position Supervision

Accurate speed control depends on reliable feedback.

Common feedback mechanisms include:

- Encoders
- Resolvers
- Magnetic floor sensors
- Limit switches

If feedback is inconsistent, the drive cannot regulate speed correctly.
Symptoms of feedback issues may include:

- Jerky motion
- Overshoot
- Speed instability
- Levelling oscillation

Diagnostic evaluation requires verification of the Engineer before adjusting drive parameters.

6.7 Common Speed Control Faults

Common motion control issues include:

- Overcurrent trips during acceleration
- Abrupt brake application
- Oscillating levelling
- Delayed deceleration

These conditions require examination of:

- Brake timing
- Drive tuning
- Feedback alignment
- Mechanical drag

Understanding the speed profile phases helps isolate where the abnormality occurs.

6.8 Transition to Landing Arrival

As the lift approaches its destination, precision becomes critical. The next chapter focuses on levelling accuracy, brake application timing, and final stopping coordination.

As shown in Figure 2.1, the next stage following levelling is:

- Brake application
- Door opening sequence

The system now transitions from motion control back to access control.

CHAPTER 7
Levelling Accuracy, Brake Application, and Final Stopping

7.1 Transition from Deceleration to Stop

At the end of deceleration, the elevator enters its most precise operational phase: final stopping and levelling.

During this stage:

- Speed is reduced to a creep velocity
- Position feedback confirms floor alignment
- Drive output decreases progressively
- Brake preparation logic activates

Stopping must occur within tight tolerance limits to ensure passenger safety and compliance with accessibility requirements.

The system does not simply cut power. It coordinates torque reduction with brake application.

7.2 Levelling Accuracy

Levelling refers to the final alignment of the car floor with the landing sill.

Acceptable tolerance is typically within a few millimetres, depending on jurisdiction and configuration.

Levelling accuracy depends on:

- Encoder resolution
- Calibrated feedback
- Drive parameter tuning
- Rope stretch compensation
- Load variation

If levelling is inaccurate, the cause is rarely random. It usually relates to feedback interpretation or mechanical wear.

Engineers should avoid adjusting parameters without confirming the integrity of the feedback.

7.3 Brake Application Timing

Brake application must be coordinated carefully.

If the brake applies too early:

- The car may stop short
- Harsh deceleration may occur
- If brake application is delayed:
- The car may drift
- Levelling may overshoot

The controller typically ensures that:

- Torque is reduced to near zero
- Motor rotation slows to a creep
- Brake coil de-energises
- The mechanical brake applies smoothly

Proper sequencing prevents shock and positional error.

Levelling and Brake Timing Diagram

FIGURE 7.1 – Levelling and Brake Timing Diagram

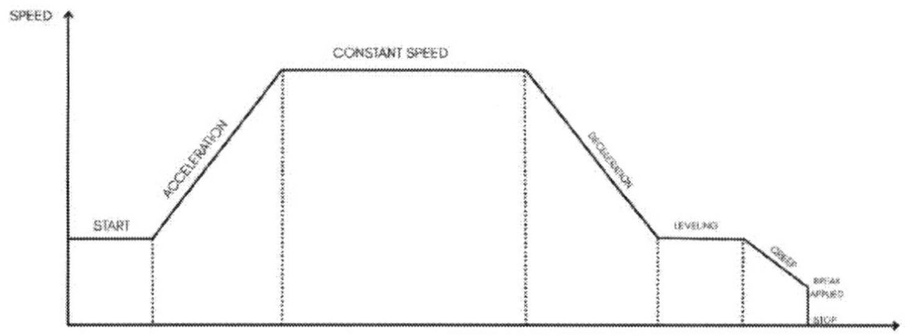

Figure 7.1 – Brake Timing and Levelling Coordination.

This diagram represents the coordinated relationship between deceleration, creep speed, and brake application during final stopping.

7.4 Drift and Post-Stop Movement

After brake application, the car should remain stationary.

If drift occurs, possible causes include:

- Weak brake coil
- Worn brake linings
- Insufficient holding torque
- Load imbalance

Drift indicates mechanical holding deficiency rather than drive instability.

Engineers should evaluate brake integrity before adjusting control parameters.

7.5 Re-levelling Behaviour

Modern systems may perform re-leveling when:

- Load changes significantly
- Rope stretch affects alignment
- Brake relaxation occurs

Re-levelling activates low-speed correction without a full travel cycle.

Frequent re-leveling may indicate:

- Mechanical wear
- Rope elongation
- Brake slip
- Feedback inconsistency

Diagnosis requires distinguishing between normal adaptive behaviour and the underlying fault.

7.6 Positional Feedback Integrity

Accurate levelling depends on reliable position data.

If position feedback drifts:

- Floors may misalign progressively
- Deceleration points shift
- Overshoot may increase

Engineers must confirm:

- Encoder mounting stability
- Signal integrity
- Calibration parameters

Feedback errors often masquerade as drive-tuning problems.

<u>Field Scenario – Case Study (Intermittent Levelling Inaccuracy)</u>

<u>Condition:</u>

The lift occasionally stops 8 to 12 mm below the landing level. Issue not consistent across all floors.

<u>Initial Assumption (Common Error):</u>

Drive deceleration parameter incorrectly set.

<u>Structured Evaluation:</u>

1. Verify encoder mounting security.
2. Check encoder coupling integrity.
3. Monitor position feedback during deceleration.
4. Inspect rope condition and tension balance.
5. Observe brake application timing.

<u>Observation:</u>

Encoder coupling slightly loose. Minor shaft slippage under deceleration load.

Feedback drift caused a deceleration reference shift.

<u>Diagnosis:</u>

Position feedback inconsistency due to mechanical encoder slippage.

Drive parameters not at fault.

<u>Corrective Action:</u>

- Secure encoder coupling
- Recalibrate positional reference
- Verify consistent levelling across multiple cycles

<u>Engineering Principle Reinforced:</u>

Feedback integrity must be verified before adjusting drive parameters.

Symptoms at the levelling stage often originate in mechanical feedback.

7.7 Diagnostic Focus

When levelling or stopping issues arise, Engineers should evaluate:

1. Brake release timing
2. Brake holding strength
3. Drive deceleration profile
4. Feedback device calibration
5. Mechanical friction or drag

Precision stopping is a coordinated system event, not a single component action.

7.8 Transition to Door Opening

As the lift approaches its destination, precision becomes critical. The next chapter focuses on levelling accuracy, brake application timing, and final stopping coordination.

As shown in Figure 2.1, the system now returns to the access control stage. The controller issues the door-open command.

The operational sequence has completed a full motion cycle and prepares to begin again.

Field Scenario – Case study (Top Terminal Overtravel Due to Brake Slide)

Condition:

Lift consistently stopping 20–30 mm above the top terminal landing.

No faults recorded.

No encoder errors logged.

No drive trip events observed.

Issue noticeable primarily on the uppermost floor.

Initial Assumption (Common Error):

Drive deceleration parameters are incorrectly tuned.

Encoder feedback drift.

Structured Evaluation:

1. Verified encoder mounting and coupling integrity.
2. Monitored deceleration profile during approach.

3. Observed motor current decay during the final stopping phase.
4. Measured brake application timing relative to zero-speed reference.
5. Performed loaded and unloaded tests.

Observation:

During upward travel to the top terminal floor:

- Deceleration occurred correctly.
- Zero speed reference achieved.
- Brake coil de-energised as expected.

However, after brake application, a slight upward drift occurred before the full mechanical holding force developed.

This drift was more pronounced under lighter load conditions.

Diagnosis:

Brake slide due to insufficient mechanical holding torque at the top terminal condition.

The drive and encoder were functioning correctly.

The stopping error occurred after propulsion was removed and before full brake torque was established.

Root Cause

Inspection revealed:

- Brake lining wear
- Slight contamination on friction surfaces
- Brake clearance slightly outside recommended tolerance

Corrective Action:

- Cleaned brake assembly
- Adjusted brake clearance
- Verified full holding torque under rated load
- Conducted repeated stopping tests at top terminal floor

Stopping accuracy restored within acceptable tolerance.

Engineering Principle Reinforced:

Not all levelling errors originate in the drive or encoder.

Stopping precision depends on:

- Accurate deceleration
- Correct brake timing
- Adequate mechanical holding torque

Terminal overtravel at the top floor is frequently associated with brake slide, particularly during upward travel, where counterweight influence increases post-stop drift potential.

Terminal-slowdown switches and final limit switches are designed to protect against excessive overtravel, but proper brake adjustment prevents the condition from developing.

<u>Field Scenario – Case study (Bottom Terminal Overtravel Due to Deceleration Fault)</u>

<u>Condition:</u>

Lift approaching the bottom terminal floor overshoots the stopping point intermittently by 40–60 mm before re-leveling upward.

Issue more noticeable under heavy load conditions.

No safety circuit interruption occurred.

No final limit activation recorded.

<u>Initial Assumption (Common Error):</u>

Brake weakness or mechanical imbalance.

<u>Structured Evaluation:</u>

1. Verified brake holding torque under rated load.
2. Inspected brake lining and clearance.
3. Monitored deceleration curve using drive diagnostics.
4. Compared encoder feedback against the commanded deceleration profile.
5. Checked the terminal slow-down switch activation timing.

<u>Observation</u>

<u>During downward travel:</u>

- Terminal slow-down sequence initiated correctly.
- Drive deceleration began at the expected position.
- However, the deceleration rate was insufficient under heavy load.

Motor current remained elevated slightly longer than expected before brake application.

The car's momentum carried it beyond the intended stopping point.

Diagnosis:

Incorrect deceleration tuning for loaded downward travel.

The brake was functioning correctly.

Encoder feedback stable.

Stopping error occurred due to insufficient braking torque during the final deceleration phase, combined with the gravitational load.

Root Cause

Drive deceleration parameter slightly under-tuned relative to rated load conditions.

Terminal slow down switch is functioning correctly.

Final limit switch not activated.

Corrective Action:

- Adjusted deceleration profile within the manufacturer's specification
- Verified braking contribution timing
- Conducted repeated loaded downward stopping tests
- Confirmed levelling accuracy restored within tolerance

Engineering Principle Reinforced

Bottom terminal overtravel under load often relates to deceleration tuning rather than brake failure.

Terminal slow-down switches initiate controlled deceleration.

Final limit switches protect against extreme overtravel.

Correct parameter tuning prevents unnecessary reliance on terminal protection devices.

CHAPTER 8

Door Opening, Call Cancellation, and Cycle Completion

8.1 From Motion to Access

Once levelling is confirmed and the brake is fully applied, propulsion ceases.

At this stage:

- The car is stationary
- Position alignment is verified
- Safety circuit remains active
- Door opening logic becomes eligible

Door opening requires verified positional alignment and supervisory continuity. The controller confirms:

- Valid floor position
- No active safety interruption
- Correct mode for passenger access

Only then is the door-open command issued.

8.2 Door Opening Sequence

The door operator now performs the reverse of the closure cycle.

The controller verifies:

- Brake fully applied
- Motor output zero
- Position confirmed
- Door zone signal active

The door-open command energizes the operator. The car and landing doors open in a coordinated sequence.

Opening must remain smooth and controlled. Abrupt door opening may indicate operator tuning issues.

FIGURE 8.1 – Door Opening Authorization Logic

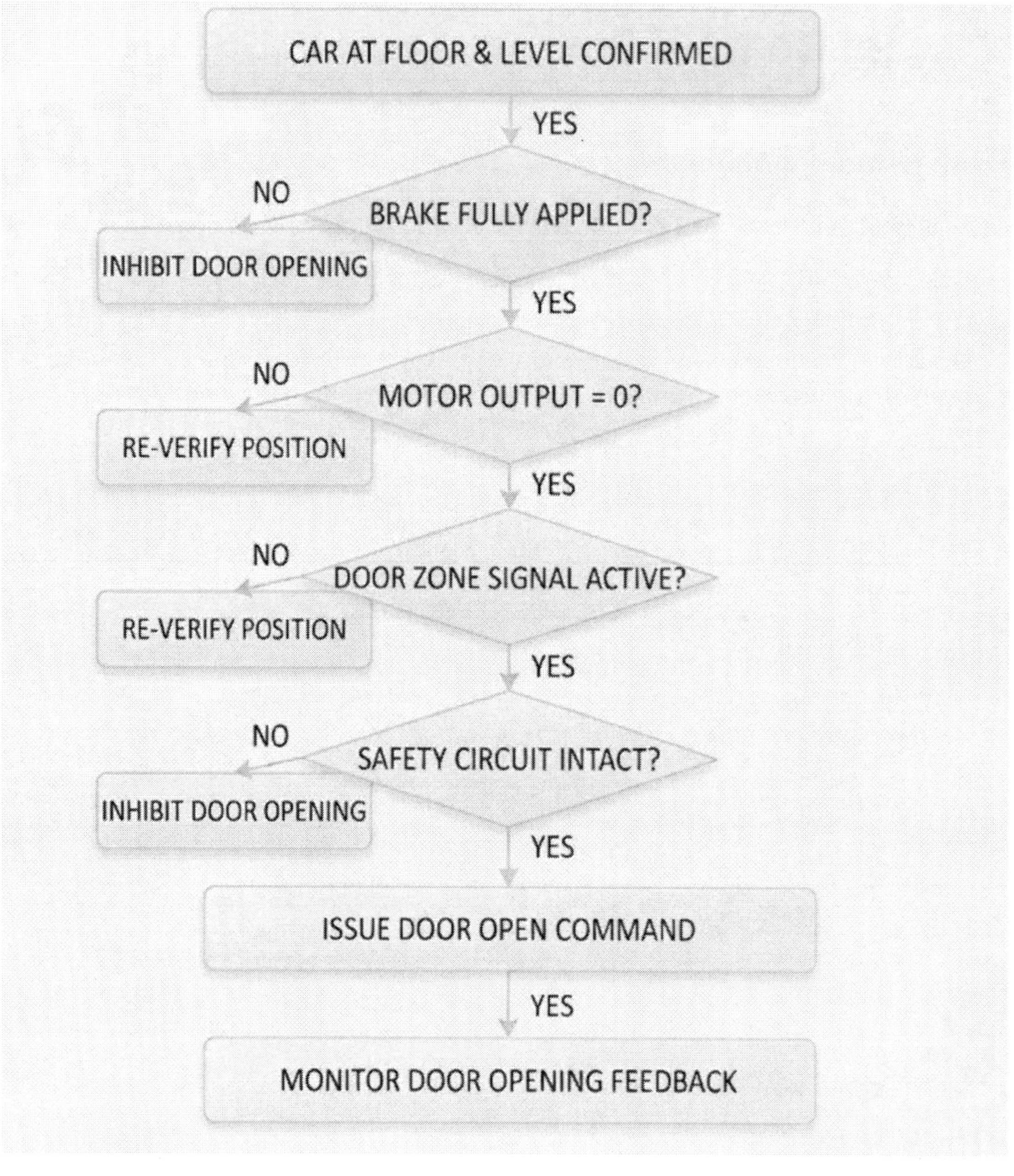

Figure 8.1 - Door Opening Authorization Logic.

This figure illustrates that door opening is authorised only after verified positional alignment, full brake application, zero propulsion output, and intact safety circuit continuity.

8.3 Call Cancellation Logic

Call cancellation occurs only after the floor has been successfully serviced.

For car calls:

- Cancellation typically occurs when doors open at the requested floor.

For landing calls:

- Cancellation occurs when the assigned car arrives and services the floor in the correct direction.

Improper call cancellation may result in:

- Repeat stops
- Phantom calls
- Apparent memory faults

Understanding the timing of cancellation is critical during diagnostics.

8.4 Door Dwell Time and Timing Control

Door dwell time is programmable.

The controller may adjust dwell based on:

- Traffic conditions
- Mode selection
- Accessibility requirements

Extended dwell in special modes is normal behaviour. Engineers must distinguish between programmed dwell and operator-delay faults.

8.5 Reopening Logic

During the open cycle, reopening may occur if:

- Obstruction detected
- Door edge activated
- Safety light curtain interrupted

Reopening logic protects passengers but may create a perception of a door malfunction if the obstruction persists.

Repeated reopening often indicates:

- Misaligned door edge
- Contaminated light curtain
- Mechanical drag

8.6 Completion of Operational Cycle

With doors fully open, the elevator has completed one full operational sequence.

The system now waits for:

- New call registration
- Timer expiration
- Mode command

The sequence may now begin again at Stage 1, as shown in Figure 2.1.

8.7 Diagnostic Focus

When investigating arrival or door opening faults, Engineers should confirm:

1. Brake fully applied
2. Position accurately confirmed
3. Door zone signal active
4. Door-open command issued
5. Feedback from the operator present

Access problems often originate in positional confirmation or operator feedback, not propulsion systems.

8.8 Transition to Special Operating Modes

Once a full operational cycle is understood, it becomes necessary to examine how alternative modes modify this sequence. The next chapter explores special operating modes and priority logic within the same supervisory framework.

CHAPTER 9
Special Operating Modes and Priority Logic

9.1 Why Special Modes Exist

Normal operation follows the sequence defined in Figure 2.1.

However, elevator systems must adapt to specific operational requirements, including:

- Maintenance access
- Emergency response
- Priority transport
- Independent control

Special operating modes alter portions of the operational sequence while preserving core safety supervision.

They do not remove the safety hierarchy.

They modify call logic, dispatch behavior, and door response.

9.2 Independent Service Mode

Independent service is typically activated using a key switch inside the car.

When active:

- Landing calls are ignored
- Only car calls are accepted
- The car does not respond automatically to external requests
- Doors remain under the car operator's control

This mode is commonly used for:

- Moving equipment
- Secured passenger transport
- Controlled movement without external interruption

Safety supervision remains intact. Door locks and the safety chain must remain in place.

9.3 Priority Mode

Priority mode allows a specific call input to override normal dispatch logic.

Examples include:

- VIP transport
- Medical priority
- Controlled service override

In this mode:

- Selected calls are serviced before standard traffic
- Other calls may be temporarily held
- Direction may be altered to satisfy the priority request

Priority modifies allocation logic but does not bypass supervision or propulsion safeguards.

FIGURE 9.1 – Mode Influence on Operational Sequence

Mode	Sequence
Normal Mode	Call → Allocation → Doors → Safety → Drive → Motion
Independent Mode	Car Call Only → Doors → Safety → Drive → Motion
Priority Mode	Priority Call Overrides → Allocation → Doors → Safety → Drive → Motion
Inspection Mode	Manual Control → Safety Supervision → Limited Motion

Figure 9.1 – Mode Influence on Operational Sequence.

This diagram shows how special operating modes modify call and dispatch logic while preserving safety circuit supervision.

9.4 Inspection Mode

Inspection mode is used during maintenance and shaft access.

When activated:

- Normal passenger calls are disabled
- Movement is controlled manually

- Speed is limited
- Certain automatic functions are bypassed

Inspection mode reduces automation but does not eliminate safety devices. Limit switches, stop circuits, and protective supervision remain active.

Engineers must confirm the inspection mode status before diagnosing an apparent call failure.

9.5 Fire Service Operation (Overview)

Fire service operation significantly alters dispatch logic.

Phase 1 recall:

- Automatically returns the car to a designated floor
- Overrides normal calls

Phase 2 operation:

- Allows manual control by authorised personnel
- Modifies door behavior

Fire service logic prioritises emergency response over traffic efficiency.

A detailed examination of the fireman's mode will be provided in a later chapter.

9.6 Mode Conflicts and Misinterpretation

Many perceived faults are actually mode-related behavior.

Examples include:

- Lift ignoring landing calls
- Doors are not responding automatically
- Car refusing normal dispatch

Engineers must confirm:

1. Current operating mode
2. Key switch status
3. Fire recall activation

9.6 Inspection controls

Mode verification should precede hardware investigation.

Field Scenario – Case Study (Lift Ignoring Landing Calls)

Condition:

Lift responds to car calls but ignores all landing calls. No visible faults present.

Initial Assumption (Common Error):

Landing call circuit failure or group controller communication fault.

Structured Evaluation

1. Verify inspection mode not active.
2. Confirm fire service not engaged.
3. Check independent service key switch position.
4. Observe landing call input at controller.

Observation:

Independent service mode active via internal key switch.

System functioning correctly under selected mode.

Diagnosis:

No hardware fault present. Mode selection altered normal dispatch logic.

Corrective Action:

- Deactivate independent service
- Confirm restoration of normal call response

Engineering Principle Reinforced:

Mode verification must precede hardware investigation.

Logic modification is not system failure.

9.7 Diagnostic Focus

When behaviour deviates from expected normal sequence:

- Confirm mode selection
- Confirm override logic

- Confirm call eligibility
- Confirm supervisory integrity

Special modes modify call and dispatch logic, but never eliminate safety supervision.

9.8 Transition to Group Control and Traffic Strategy

Beyond individual car logic, multi-lift installations introduce additional allocation complexity. The next chapter examines group control strategies, dispatch logic, and traffic optimisation.

CHAPTER 10
Group Control, Dispatch Strategy, and Traffic Management

10.1 From Simplex to Group Operation

In a simplex installation, one lift responds to all calls.

The logic flow remains straightforward:

Call → Allocation (implicit) → Direction → Door → Safety → Drive → Motion

In group systems, however, call allocation becomes an active decision stage.

Two or more lifts must:

- Share landing calls
- Avoid duplication
- Minimise passenger waiting time
- Maintain directional efficiency

Group control introduces a higher-level decision layer before propulsion logic begins.

10.2 Collective Control Types

Collective control determines how calls are handled.

<u>Up Collective</u>

- Lift answers all upward calls while travelling upward
- Downward calls are ignored until direction changes

<u>Down Collective</u>

- Lift answers all downward calls while travelling downward
- Upward calls are deferred

<u>Full Collective</u>

- Lift responds to both up and down calls in sequence
- Direction may change after completing commitments

FIGURE 10.1 – Collective Control Comparison

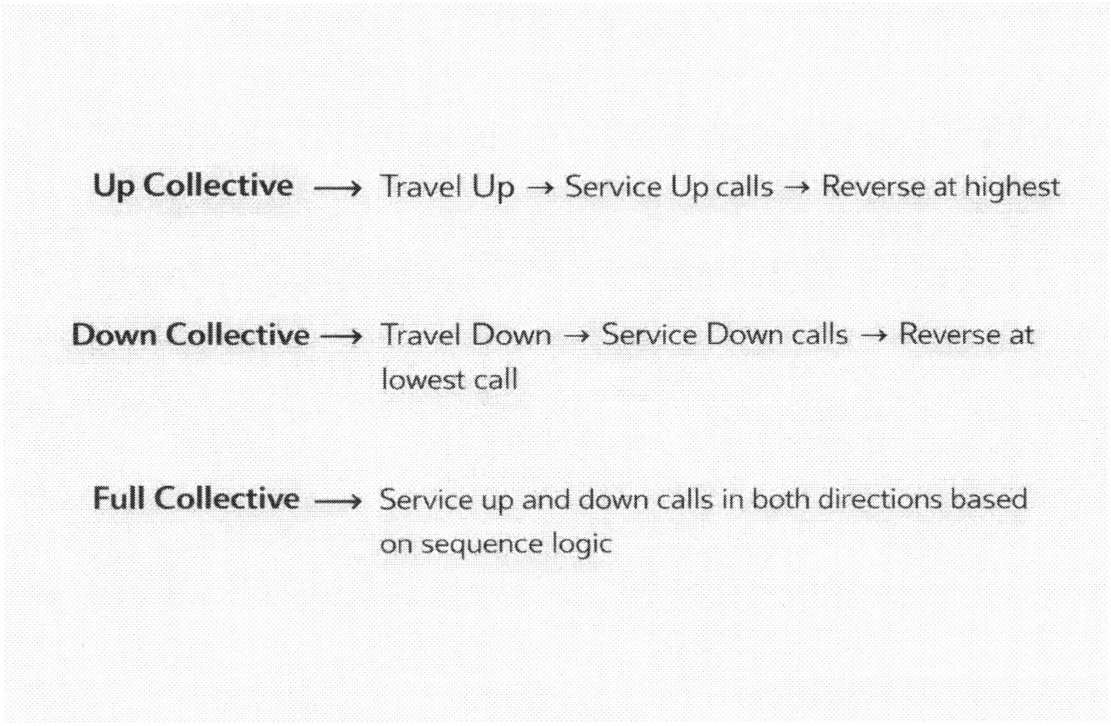

Figure 10.1 – Collective Control Strategies.

This figure illustrates the operational differences between up collective, down collective, and full collective control methods.

10.3 Duplex and Triplex Operation

In duplex systems:

- Two cars share landing calls
- The allocation logic selects the most appropriate car

In triplex or larger systems:

- Allocation becomes dynamic
- Position, load, and direction are evaluated continuously

The objective is not merely a response, but efficiency.

10.4 Allocation Logic

When a landing call is registered in a group system, the controller evaluates:

- Current car positions
- Direction of travel

- Estimated time to arrival
- Load conditions
- Existing commitments

The selected car receives the assignment.

FIGURE 10.2 – Simplified Group Allocation Logic

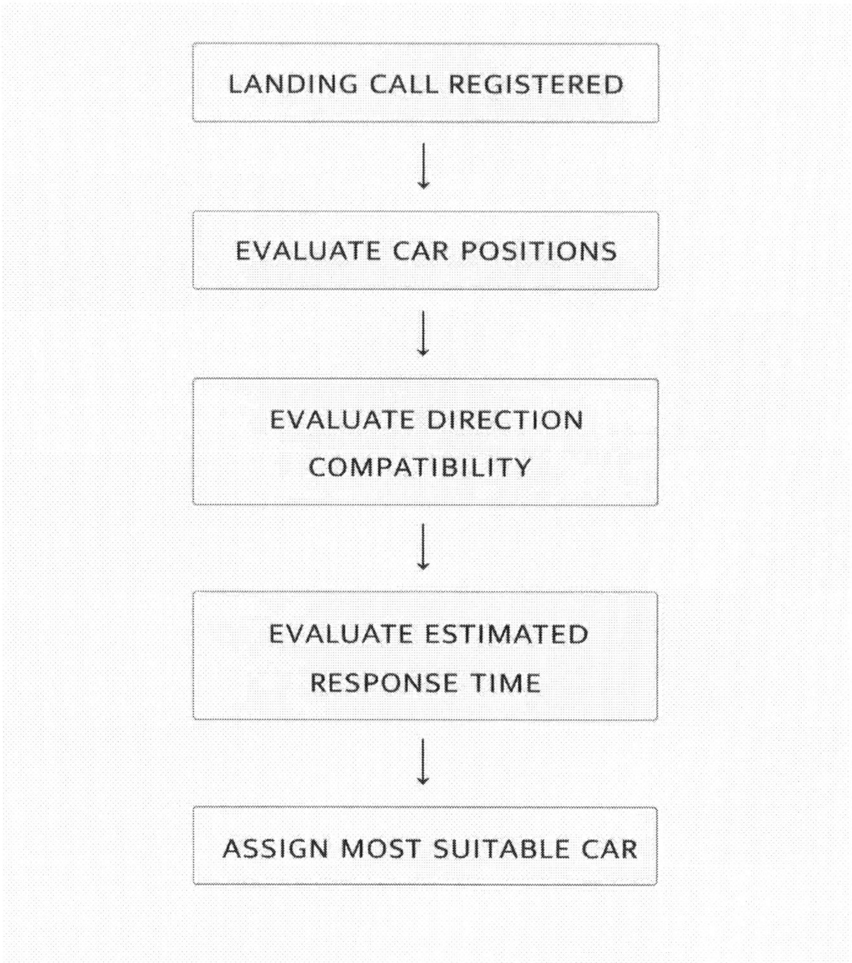

Figure 10.2 – Group Allocation Logic Model.

This diagram represents the simplified decision process used by the group controller to assign landing calls to the most appropriate car.

10.5 Traffic Patterns

Traffic patterns influence dispatch logic.

Common patterns include:

- Morning up-peak

- Lunchtime mixed flow
- Evening down-peak

Advanced systems may adapt dynamically to traffic patterns to reduce waiting time.

Understanding these patterns helps Engineers interpret system behaviour during busy periods.

10.6 Destination Control Systems

In destination control:

- Passengers select their destination before entering the car
- The system assigns a specific car
- Stops are optimised to reduce unnecessary travel

Destination control reduces intermediate stops and improves traffic efficiency.

However, the underlying supervision and propulsion sequence remains unchanged.

10.7 Diagnostic Focus

When group behaviour appears irregular, Engineers should confirm:

1. Correct car allocation
2. Proper direction handling
3. Mode restrictions
4. No conflicting overrides
5. Communication integrity between cars

Group faults are often communication- or configuration-related rather than mechanical.

10.8 Transition to System Integrity

While group logic determines which car moves, system integrity determines whether movement is permitted.

Group coordination depends on reliable communication and signal integrity. The next chapter examines control signals, feedback devices, and position supervision within closed-loop architecture.

CHAPTER 11
Control Signals, Feedback Devices, and Position Supervision

11.1 Command vs Feedback

Elevator systems operate on two fundamental signal categories:

- Command signals
- Feedback signals

Command signals originate from the controller. They instruct subsystems to perform actions.

Feedback signals originate from devices in the field. They confirm whether commanded actions have been achieved.

Reliable operation depends on alignment between command and feedback. If a command is issued but feedback is not received, the system halts progression.

11.2 Control Signal Pathways

Control signals typically include:

- Door open and close commands
- Drive enable signals
- Brake release commands
- Direction selection signals

These signals travel from the controller to output relays or drive inputs.

Their integrity depends on:

- Proper voltage levels
- Secure terminations
- Relay contact condition
- Wiring continuity

A missing command may appear to be a subsystem failure when the actual issue lies upstream.

FIGURE 11.1 – Control and Feedback Architecture

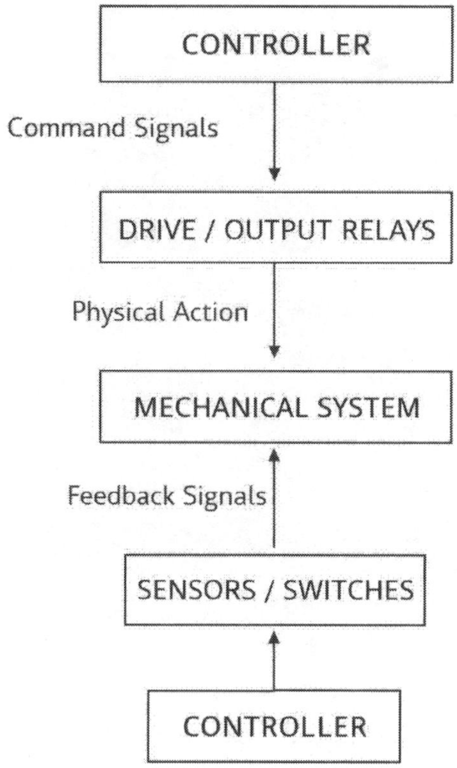

Figure 11.1 - Closed-Loop Control Architecture.

This figure illustrates the relationship between the controller command signals and the field feedback devices in a closed-loop supervisory system.

11.3 Position Feedback Devices

Accurate motion and levelling require position supervision.

Common position feedback devices include:

- Rotary encoders
- Resolvers
- Magnetic vane sensors
- Proximity switches
- Limit switches

Encoders provide continuous positional data, allowing precise speed and deceleration control.

Limit switches provide boundary protection.

Door zone sensors confirm arrival at the landing level.

11.4 Encoder Integrity

Encoders play a central role in speed and position regulation.

Fault symptoms may include:

- Erratic levelling
- Sudden deceleration
- Overshoot
- Drive overcurrent trips

Encoder issues may arise from:

- Loose mounting
- Shaft slippage
- Electrical noise
- Damaged signal cable

Before adjusting drive parameters, feedback integrity must be verified.

11.5 Limit and Final Limit Switches

Limit switches prevent overtravel.

Final limit switches provide the ultimate safety cut-off.

These devices are not part of routine dispatch logic but form part of the supervisory hierarchy.

If activated:

- Safety chain opens
- Drive output ceases
- Brake applies

Engineers must investigate the root cause before resetting.

11.6 Door Zone Signals

Door zone signals confirm that the car is within an acceptable alignment range for door opening.

If the door zone signal is absent:

- Door open command is inhibited
- Re-levelling may be triggered

Faulty door zone sensors can create apparent levelling issues even when positioning is correct.

11.7 Communication Signals in Group Systems

In group installations, cars communicate with each other and with the group controller.

Signal loss may result in:

- Incorrect allocation
- Duplicate dispatch
- Unresponsive landing calls

Network integrity must be considered when diagnosing group faults.

11.8 Diagnostic Focus

When diagnosing control or feedback anomalies, Engineers should determine:

1. Was a command issued?
2. Was feedback received?
3. Does feedback match the commanded state?
4. Is the signal voltage stable?
5. Is noise or interference present?

Control systems depend on reliable closed-loop confirmation.

11.9 Transition to Power Integrity

Signal integrity depends on a stable electrical supply and protective grounding. The next chapter examines power distribution, earthing, and protection mechanisms that support reliable operation.

CHAPTER 12
Electrical Power Supply, Earthing, and Protection

12.1 Power as the Foundation of Control

All command, supervision, and propulsion functions depend on a stable electrical supply.

An elevator typically uses:

- Three-phase power for propulsion
- Single-phase or derived supply for control circuits
- Low-voltage supply for logic and sensors

Without stable voltage and proper distribution, signal integrity and propulsion cannot be trusted.

Before diagnosing logic faults, Engineers must confirm power integrity.

12.2 Power Distribution Structure

Power distribution in a typical installation includes:

- Main incoming supply
- Main isolator
- Circuit protection devices
- Control transformer
- Drive supply input
- Auxiliary power circuits

The control and propulsion systems may share a common supply origin, but are often separated for stability and protection.

FIGURE 12.1 – Simplified Elevator Power Architecture

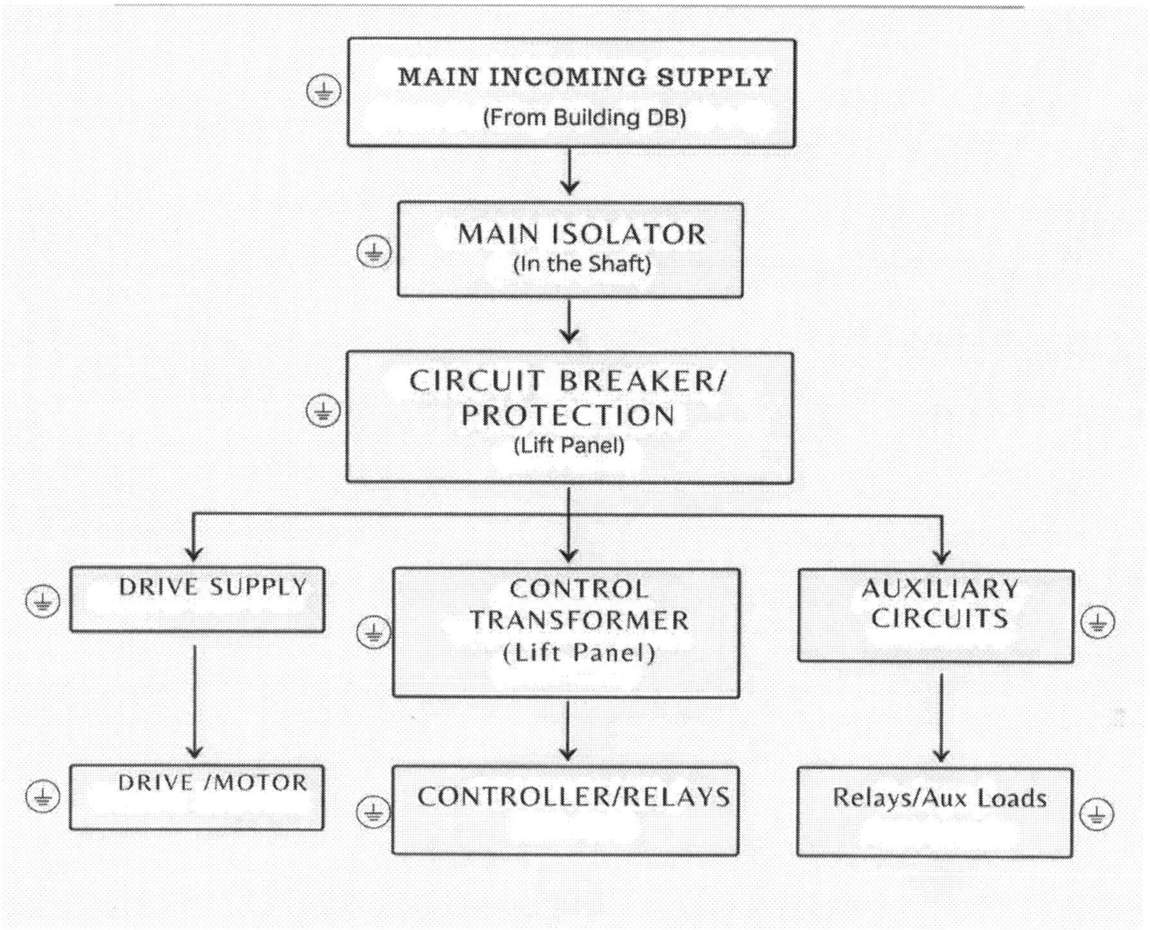

Figure 12.1 – Elevator Power Distribution Architecture.

This schematic shows the distribution of incoming power to propulsion, control, and auxiliary circuits in a typical installation.

12.3 Control Voltage Stability

Control circuits typically operate at a lower voltage.

Control voltage instability may cause:

- Relays dropping intermittently
- Contactors chattering
- False fault logging
- Unexpected resets

High resistance in control circuits, as seen in the previous case study, can reduce effective voltage under load.

Voltage measurement under operating conditions is essential.

12.4 Three-Phase Supply Integrity

Propulsion systems depend on a balanced three-phase supply.
Common supply-related issues include:

- Phase loss
- Voltage imbalance
- Loose terminals
- Poor isolation

Phase imbalance can cause:

- Motor overheating
- Drive overcurrent trips
- Reduced torque

Drive fault logs may reflect supply conditions rather than internal failure.

12.5 Earthing and Ground Integrity

Proper earthing serves two purposes:

- Safety for personnel
- Stability for electronic control systems

Poor grounding may result in:

- Electrical noise
- Interference in encoder signals
- Communication instability
- Unpredictable controller behaviour

Ground integrity should be verified during commissioning and fault investigation.

12.6 Protection Devices

Protection systems include:

- Circuit breakers
- Fuses
- Overcurrent protection
- Thermal protection
- Surge protection

These devices protect equipment from catastrophic failure.

Frequent tripping indicates underlying electrical stress.

Engineers must identify the cause rather than simply resetting protection devices.

12.7 Emergency Power and Backup Systems

Some installations include:

- Emergency lighting
- Battery-backed lowering systems
- Generator supply

Emergency power may alter behaviour during supply loss.

Understanding backup logic prevents misdiagnosis during power interruptions.

12.8 Diagnostic Focus

When diagnosing electrical issues, Engineers should evaluate:

1. Incoming voltage stability
2. Phase balance
3. Control voltage under load
4. Earthing integrity
5. Terminal tightness
6. Protection device status

Electrical stability forms the foundation of reliable control logic.

12.9 Transition to Commissioning and Verification

With operational and electrical integrity understood, verification becomes essential. The next chapter examines commissioning logic and structured testing procedures.

CHAPTER 13
Commissioning Logic, Testing, and Verification

13.1 Commissioning as Proof, Not Procedure

Commissioning is not the act of making a lift move.

It is the structured verification that every required supervisory, control, and propulsion condition operates correctly under real conditions.

A moving lift is not necessarily a safe lift.

A commissioned lift has been verified logically, sequentially, and systematically.

Commissioning follows the same operational order described earlier in this book. Each stage is validated independently before the next stage is introduced.

13.2 Pre-Motion Verification

Before any propulsion testing begins, Engineers must confirm:

- Power supply integrity
- Correct phase sequence
- Control voltage stability
- Safety circuit continuity
- Door lock verification
- Correct wiring polarity

This stage confirms that the system is electrically stable and safe to test.

No motion testing should occur until supervisory integrity is confirmed.

FIGURE 13.1 – Structured Commissioning Sequence

Figure 13.1 – Structured Commissioning Sequence.

This figure illustrates the recommended logical order for commissioning and verifying elevator system operation.

13.3 Incremental Motion Testing

Initial movement should occur at low speed.

During the first motion:

- Confirm brake release timing
- Observe smooth acceleration
- Monitor current draw
- Verify safety relay stability

High-speed testing should only follow successful low-speed validation.

Skipping stages increases risk.

13.4 Safety Device Testing

Every protective device must be individually tested.

This includes:

- Door lock interruption
- Emergency stop activation
- Inspection switch function
- Overspeed simulation
- Buffer switches
- Final limit verification

Safety devices are not assumed to function. They must be proven.

Testing confirms that supervision governs propulsion under fault conditions.

13.5 Brake Verification

Brake testing includes:

- Static holding test
- One-sided brake test
- Loaded holding verification
- Controlled stopping test
- Drift observation

Brake integrity directly affects passenger safety. Commissioning must verify holding capacity under rated load.

13.6 Levelling Accuracy Testing

Levelling should be verified:

- At multiple floors
- Under varying load conditions
- During repeated cycles

Consistency matters more than isolated accuracy. Engineers should monitor:

- Positional stability

- Repeatability
- Re-levelling behaviour

13.7 Group System Verification

In group installations:

- Confirm the correct car allocation
- Verify no duplicated dispatch
- Confirm communication integrity
- Observe traffic handling under simulated peak load

Group logic must be validated before handover.

13.8 Documentation and Record Keeping

Professional commissioning requires documentation.

Records should include:

- Voltage readings
- Safety test results
- Brake performance
- Levelling tolerance
- Fault log review

Documentation protects both the engineer and client.

13.9 Commissioning Discipline

Commissioning demands patience and method.

Shortcuts often create latent faults that appear after handover.

A structured sequence ensures:

- Safety devices operate correctly
- Propulsion functions predictably
- Control logic behaves consistently

Commissioning confirms that the theoretical sequence described in this book functions correctly in practice.

13.10 Transition to Diagnostics Under Fault Conditions

After commissioning confirms proper operation, real-world conditions introduce wear and unexpected behaviour.

The next chapter examines structured diagnostics and sequence-based troubleshooting.

CHAPTER 14
Advanced Diagnostics and Structured Troubleshooting

14.1 Symptoms Are Not Causes

In live environments, Engineers are often presented with symptoms:

- Lift stuck between floors
- Doors not opening
- Intermittent stopping
- Random drive fault codes

Symptoms describe behaviour, not the cause.

Effective diagnosis requires tracing the symptom backward through the operational sequence described in earlier chapters.

The key question becomes: At which stage did progression stop?

14.2 Sequence-Based Fault Isolation

Using Figure 2.1 as a reference, Engineers can identify the final verified stage in the sequence.

Example:

- Call registered
- Direction established
- Doors closed
- No movement

This suggests an investigation at:

- Safety circuit continuity
- Drive enable
- Brake release

By locating the halted stage, the search area narrows significantly.

14.3 Structured Diagnostic Approach

A disciplined troubleshooting process should follow five steps:

1. Confirm the symptom precisely
2. Identify the last confirmed operational stage
3. Verify required inputs for the next stage
4. Measure signal integrity under load
5. Confirm that corrective action resolves the progression

This structure prevents random component replacement.

FIGURE 14.1 – Structured Diagnostic Framework

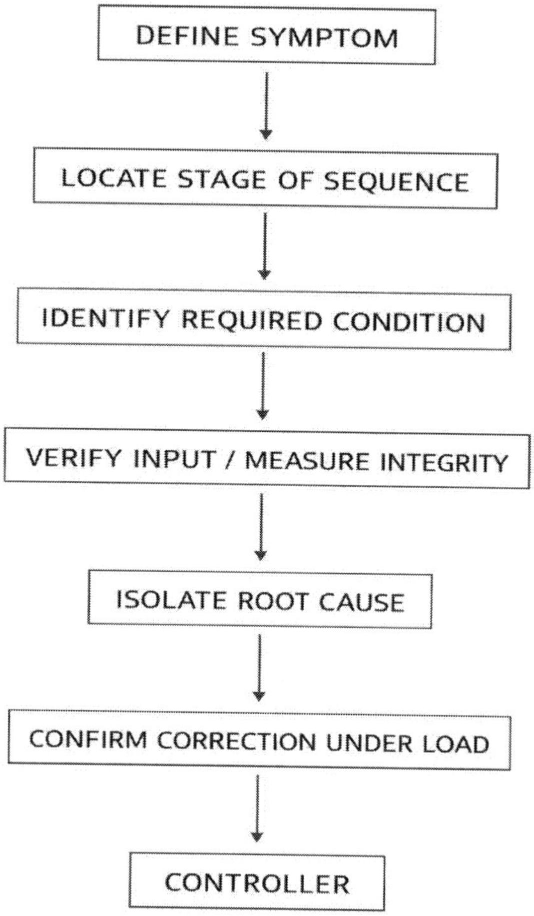

Figure 14.1 - Structured Diagnostic Framework.

This diagram represents the sequence-based troubleshooting model used to isolate faults by identifying the halted operational stage.

14.4 Open Circuit vs High Resistance

Not all circuit failures are binary. An open circuit prevents progression completely. A high-resistance condition may allow partial function before failure.

As demonstrated in the earlier safety circuit case study, high resistance can reduce coil voltage without registering a discrete fault.

Engineers must measure voltage under load rather than relying solely on continuity checks.

14.5 Intermittent Faults

Intermittent faults require:

- Observation during motion
- Vibration simulation
- Load testing
- Thermal inspection

Replacing components without isolating the cause can temporarily mask the issue.

Structured reasoning reveals the unstable condition.

14.6 Fault Logs and Interpretation

Error logs record detected anomalies.

They do not always identify the root cause.

Logs may indicate:

- Drive trip
- Safety interruption
- Communication loss

Engineers must determine whether the logged event is primary or secondary.

The system records what caused it to stop, not necessarily what caused the stoppage.

14.7 Avoiding Diagnostic Bias

Common diagnostic errors include:

- Replacing the component named in the fault log
- Assuming recent work is unrelated
- Ignoring intermittent mechanical causes
- Adjusting parameters prematurely

Bias reduces diagnostic clarity. Engineers must remain methodical.

14.8 Escalation and Documentation

When faults persist:

- Document measurements
- Record sequence observations
- Confirm voltage levels
- Log environmental factors

Clear documentation supports escalation to manufacturers or supervisors when required.

Professional troubleshooting includes traceability.

14.9 Transition to Human Factors and Professional Responsibility

Technical systems operate within human environments.

The next chapter examines how fatigue, cognitive bias, and pressure influence engineering judgment.

CHAPTER 15
Human Factors, Fatigue, and Cognitive Bias in Engineering Practice

15.1 Engineering Does Not Occur in Isolation

Elevator systems follow structured logic. Human judgment does not follow structured logic automatically. Human judgment operates under pressure, fatigue, routine familiarity, and expectation. Even the most technically sound systems depend on disciplined human oversight. Many faults are not caused by system failure, but by premature assumption or incomplete verification. Understanding human factors is therefore as important as understanding supervision circuits.

15.2 Fatigue and Decision Degradation

Fatigue affects:

- Reaction time
- Risk perception
- Attention to detail
- Willingness to re-check measurements

In live environments, fatigue may result from:

- Extended callouts
- Emergency breakdowns
- Repeated entrapments
- Long commissioning sessions

Under fatigue, Engineers may:

- Skip verification steps
- Trust assumptions
- Accept incomplete confirmation

A structured process exists to protect against human variability.

15.3 Routine Familiarity Risk

Familiar systems create comfort. Comfort reduces vigilance. When Engineers service the same installation repeatedly, they may assume:

- Wiring is unchanged
- Previous repairs remain intact
- Adjustments are stable

This familiarity can mask gradual degradation. Routine installations require the same structured verification as unfamiliar ones.

15.4 Confirmation Bias

Confirmation bias occurs when Engineers interpret evidence in a way that supports an initial assumption.

Example:

- Fault log indicates drive trip
- Engineer assumes drive failure
- Replaces drive
- The underlying safety circuit issue remains

The earlier case study clearly demonstrated this principle. Structured reasoning counters bias by forcing the evaluation of required conditions before accepting conclusions.

15.5 Authority and Assumption

In some environments, prior advice influences diagnosis. If a component has already been replaced, Engineers may hesitate to question the decision. Professional integrity requires independent verification. Systems do not recognise hierarchy. They respond only to conditions.

15.6 Pressure and Escalation

Passenger entrapments create urgency. Management pressure increases the expectation for immediate resolution.

Under pressure, there is temptation to:

- Reset repeatedly
- Replace major components
- Accept an incomplete explanation

Pressure should accelerate structured reasoning, not replace it.

15.7 Documentation as Protection

Recording:

- Measured voltages
- Observed sequence
- Fault timing
- Environmental conditions

Protects both the Engineer and the client; documentation forces clarity and transforms opinion into recorded fact.

15.8 Structured Thinking as Safeguard

The sequence logic presented earlier in this book is not only technical. It protects against cognitive error.

When Engineers ask:

Which condition has not been satisfied?

They are resisting bias. They are grounding judgment in structure.

15.9 Professional Maturity

Technical skill can be developed through training.

Professional maturity develops through:

- Reflection
- Accountability
- Willingness to admit uncertainty
- Refusal to speculate

Elevator engineering occurs in safety-critical environments. Discipline must extend beyond hardware.

15.10 Transition to Ethical Responsibility

Understanding human influence leads directly to professional responsibility.

The next chapter examines ethics, accountability, and the standards expected in safety-critical systems.

CHAPTER 16
Ethics, Accountability, and Professional Responsibility

16.1 Engineering as Public Trust

Elevator systems operate in hospitals, schools, residential buildings, commercial towers, and transport hubs. The public uses them without questioning the safeguards behind their operation.

That trust is not abstract. It is maintained by Engineers who ensure that supervision, propulsion, and protection systems function as intended.

Engineering decisions directly affect safety.

Professional responsibility, therefore, extends beyond technical correctness. It includes disciplined judgment and refusal to compromise integrity.

16.2 Accountability in Safety-Critical Systems

In safety-critical environments, small oversights can have disproportionate consequences.

Examples include:

- Failing to verify safety chain continuity
- Adjusting parameters without confirming feedback integrity
- Resetting faults without identifying the root cause
- Leaving intermittent conditions unresolved

Accountability means confirming that conditions are satisfied before restoring service.

It also means recognising when escalation is necessary.

16.3 Ethical Boundaries

Engineers may encounter situations where:

- Time pressure encourages shortcuts
- Budget constraints limit repair scope
- Clients request a temporary bypass
- Management prioritises uptime

Professional ethics require that safety supervision never be compromised.

Bypassing protective devices or ignoring high-resistance conditions may temporarily restore motion, but it undermines system integrity.

Ethical discipline often requires refusing unsafe requests.

16.4 Documentation and Transparency

Clear documentation protects:

- Engineers
- Clients
- Passengers

Accurate reporting should include:

- Observed condition
- Measured values
- Recommended corrective action
- Any limitations or risks

Transparency builds long-term credibility.

16.5 Continuous Learning

Elevator systems evolve. Drive technology advances. Control architectures modernise. Regulations update. Professional responsibility includes continuous learning and a willingness to revisit assumptions. Structured thinking, as outlined throughout this book, supports adaptability.

16.6 Mentorship and Knowledge Transfer

Experience carries responsibility.

Senior Engineers should:

- Share structured diagnostic methods
- Demonstrate safe commissioning practice
- Encourage documentation discipline
- Model ethical refusal when necessary

Knowledge transfer strengthens industry standards.

16.7 Responsibility Under Pressure

During entrapments or emergency scenarios, urgency is high.

Responsibility includes:

- Maintaining calm reasoning
- Avoiding speculation
- Communicating clearly
- Documenting accurately

Pressure does not change sequence logic.

Supervision and verification remain essential.

16.8 The Engineer's Signature

Every installation, commissioning, or corrective action carries the signature of the responsible Engineer.

Even when not physically signed, the outcome reflects professional judgement.

Integrity is demonstrated through:

- Structured reasoning
- Respect for the supervision hierarchy
- Careful verification
- Honest reporting

16.9 Transition to Closing Reflections

Professional responsibility extends beyond immediate repair.

The next chapter examines long-term discipline and the lasting impact of engineering decisions.

CHAPTER 17
Professional Legacy and Long-Term Engineering Discipline

17.1 Beyond Immediate Repair

Most engineering work is judged in the short term:

- The lift moves again.
- The fault clears.
- The callout is closed.

However, the true measure of engineering discipline is not whether motion is restored, but whether system integrity is preserved.

Each decision leaves a technical footprint.

17.2 The Long View

Elevator systems operate for decades.

A temporary adjustment made today may:

- Affect levelling accuracy years later
- Influence wear patterns
- Mask a developing supervisory issue
- Create hidden electrical stress

Long-term thinking distinguishes repair from correction.

Corrective action restores sequence integrity, not just function.

17.3 Structured Thinking as a Habit

Throughout this book, sequence-based reasoning has been emphasised.

This approach should become habitual.

When faced with any fault, Engineers should instinctively ask:

- Where did the sequence stop?
- Which condition is missing?
- What evidence confirms this?

Over time, this discipline becomes automatic.

17.4 Influence Through Example

Professional legacy is often quiet.

Junior Engineers observe:

- How faults are approached
- Whether assumptions are challenged
- Whether documentation is thorough
- Whether shortcuts are refused

Consistently demonstrated structured discipline influences industry standards more than policy documents.

17.5 Integrity Under Constraint

Constraints will always exist:

- Budget limitations
- Time pressure
- Operational urgency
- Client expectation

Legacy is built by maintaining supervisory integrity under those constraints.

A lift that operates decades after installation safely reflects disciplined engineering decisions made long before.

17.6 The System Remembers

Systems record:

- Fault logs
- Parameter changes
- Component replacements

They also reflect unseen decisions:

- Proper torque values
- Correct alignment
- Cleaned contacts
- Secured terminations

Professional work leaves fewer hidden vulnerabilities.

17.7 Contribution to Industry Standards

Engineering practice evolves collectively.

Every disciplined intervention contributes to:

- Safer installations
- More reliable operation
- Stronger diagnostic culture

When structured reasoning becomes a normal practice, industry reliability improves.

17.8 Transition to Final Integration

With a professional legacy established, the final chapter consolidates the complete operational framework presented throughout this book.

CHAPTER 18
Integrated System Reflection and Final Summary

18.1 The Complete Operational Framework

This book has followed the operational sequence of an elevator system from its first input to its final stop.

The journey began with call registration and progressed through:

- Direction determination
- Door verification
- Safety circuit confirmation
- Drive enable
- Brake release
- Controlled motion
- Deceleration and levelling
- Door opening
- Mode variation
- Group coordination
- Signal supervision
- Power integrity
- Commissioning discipline
- Structured diagnostics
- Human factors and responsibility

Each stage depends on the integrity of the stage before it. The system advances only when the required conditions are satisfied.

18.2 Sequence as a Diagnostic Map

The operational sequence is more than a description of movement. It is a diagnostic map. When a lift fails to perform as expected, the Engineer can locate the halted stage and evaluate the conditions required for progression.

This method reduces speculation. It narrows the investigation. It reinforces disciplined reasoning.

18.3 Supervision as Structural Protection

Supervision governs propulsion. From door locks to final limits, protective devices are embedded within the operational chain.

They do not exist as optional safeguards. They define eligibility for motion. Understanding this hierarchy transforms troubleshooting into structured verification rather than reactive adjustment.

18.4 Closed-Loop Integrity

Command and feedback operate in a continuous relationship. Power stability supports control logic. Control logic authorises propulsion. Feedback confirms position and movement. Supervision remains active throughout.

This closed-loop architecture ensures reliability when properly maintained. When instability appears, the source is almost always found within one of these relationships.

18.5 Commissioning and Verification Discipline

Commissioning is proof that the theoretical sequence functions correctly in physical installation.

Verification protects against hidden weaknesses. Documentation protects professional integrity. A moving system is not necessarily a proven one.

18.6 Professional Responsibility

Technical clarity alone is insufficient. Human factors influence interpretation. Bias influences assumptions. Pressure influences decision-making. Structured sequence logic provides stability under these conditions. Professional responsibility requires adherence to this discipline.

18.7 The Engineer's Framework

By the conclusion of this book, the Engineer should possess a structured framework:

1. Understand the operational sequence
2. Identify the halted stage
3. Verify required conditions
4. Confirm signal integrity
5. Evaluate the supervision hierarchy
6. Document findings
7. Confirm corrective action under load

This framework applies regardless of system brand or configuration. It is transferable because it is based on principles rather than specific products.

18.8 Closing Perspective

Elevators operate quietly within buildings, yet they represent complex coordination between mechanical systems, electrical supervision, control logic, and human judgment.

When each stage is clearly understood and systematically verified, reliability becomes predictable rather than accidental.

The sequence described in this book is not theoretical. It reflects how systems are designed to protect themselves and their users.

Clarity of reasoning strengthens safety. Structured verification strengthens reliability. Professional discipline strengthens industry standards. The operational sequence is complete.

GLOSSARY

Key Terms and Definitions

Acceleration Phase

The controlled increase in motor speed following brake release, managed by the drive to ensure smooth transition from standstill to constant travel speed.

Allocation (Implicit)

The automatic assignment of a call to a lift in a simplex system where only one car is present. In such systems, allocation does not require group decision logic, as the single car inherently receives all valid calls.

Implicit allocation differs from group allocation, where multiple cars are evaluated before assignment.

Brake Coil

An electrically energised component that releases a spring-applied mechanical brake, allowing the motor to rotate.

Brake Slide

A condition in which the mechanical brake does not apply sufficient holding force after propulsion is removed, allowing the car to continue moving slightly beyond its intended stopping point.

Brake slide may result from worn brake linings, contamination, incorrect adjustment, or insufficient brake torque.

Brake slide is a common contributor to terminal overtravel conditions.

Buffer

An energy-absorbing safety device installed at the bottom of the lift shaft to protect against excessive downward overtravel. Buffers are designed to absorb kinetic energy if the car or counterweight travels beyond its normal stopping limits.

Buffers may be spring-type or hydraulic-type, depending on rated speed and installation requirements.

Buffer Switch

A safety device associated with the buffer assembly that monitors buffer compression or engagement.

If the buffer is compressed beyond its normal position, the buffer switch interrupts the safety circuit and prevents further motion until the condition is investigated and reset.

Buffer switches ensure that the system cannot resume operation after buffer impact without proper verification.

Closed-Loop Control

A control method in which system output is continuously monitored through feedback devices to confirm that commanded actions have been achieved.

Collective Control

A call-handling method in which a lift services calls in a particular direction before reversing travel direction.

Commissioning

The structured process of verifying that all supervisory, control, and propulsion systems operate correctly and safely prior to handover.

Control Voltage

Low-voltage electrical supply used for relays, logic circuits, and supervisory devices, distinct from propulsion power supply.

Diagram

A simplified visual representation used to illustrate relationships, logic flow, system structure, or component interaction.

Diagrams in this publication are conceptual and intended to support understanding of operational principles rather than replicate manufacturer-specific wiring layouts.

Deceleration Phase

The controlled reduction in motor speed as the lift approaches its destination floor.

Destination Control

A dispatch system in which passengers select their destination floor before entering the lift, allowing optimized car assignment.

Door Interlock

A mechanical and electrical device that ensures landing doors are fully closed and locked before motion is permitted.

Door Zone

A positional range within which the lift is considered correctly aligned with a landing, allowing door opening.

Drive Enable

A control signal authorizing the inverter or motor control system to prepare for torque production.

Encoder

A feedback device that provides continuous position and speed information to the control system.

Error

A detected deviation between expected system behaviour and actual operational condition.

Errors may be logged by the controller or drive; however, an error code typically reflects the symptom of interruption rather than the underlying root cause.

Figure

A numbered visual reference within the text, including diagrams, schematics, logic models, or graphical representations used to clarify technical concepts.

Figures are sequentially numbered and referenced within the relevant chapter.

Final Limit Switch

A secondary and ultimate protective device located at the extreme terminal floors. If activated, it interrupts the safety circuit and immediately removes propulsion authority.

See also: Terminal Floors, Overtravel, Terminal Slow Down Switches.

Full Collective

A call-handling method that allows servicing of both upward and downward calls in sequence based on system logic.

High Resistance Fault

A circuit condition in which electrical continuity exists but resistance is elevated, potentially reducing voltage under load and causing intermittent malfunction.

Independent Service Mode

An operating mode in which landing calls are ignored and only car calls are accepted.

Inspection Mode

A maintenance operating mode that disables normal passenger calls and allows controlled manual movement at reduced speed.

Levelling

The final alignment of the car floor with the landing floor to within specified tolerance.

Motion Permission

The cumulative result of verified supervisory conditions that allow propulsion to occur.

Overspeed Governor

A mechanical safety device that detects excessive speed and interrupts propulsion if triggered.

Overtravel (also referred to as Overrun)

The distance a lift car travels beyond the intended stopping point at terminal floors.

See also: Brake Slide, Final Limit Switch.

Phase Imbalance

A condition in which the three-phase power supply is not evenly distributed, potentially affecting motor performance.

Position Supervision

The monitoring of car location using feedback devices to ensure safe stopping and door operation.

Priority Mode

An operating mode that allows selected calls to override normal dispatch logic.

Propulsion

The controlled generation of torque by the motor and drive system that produces car movement.

Propulsion occurs only after motion permission has been granted through verified safety circuit continuity, confirmed door locking, valid drive enable signal, and brake release coordination.

Root Cause

The underlying technical condition responsible for producing an observed symptom or system error.

Identifying root cause requires structured evaluation of sequence progression rather than reliance on fault codes alone.

Safety Circuit

A series-connected chain of protective devices that must maintain electrical continuity for motion to be authorised.

Schematic

A technical drawing that represents electrical or logical connections using symbolic notation.

Schematics illustrate functional relationships between components and are distinct from pictorial diagrams or physical wiring layouts.

All schematics in this publication are conceptual and intended for educational purposes.

Supervision

The continuous monitoring of protective and operational conditions required for safe lift movement.

Technical Findings

Documented measurable observations obtained during structured diagnostic evaluation.

Technical findings are based on evidence such as voltage readings, signal integrity verification, mechanical inspection, and system behaviour under load.

Technical findings support identification of root cause and validation of corrective action

Terminal Floors

The uppermost and lowermost floors of the lift shaft that define the designed limits of normal car travel. Protective devices such as terminal slow down switches and final limit switches are positioned at or near the terminal floors to prevent overtravel.

Terminal Slow Down Switches

Protective devices installed near the terminal floors that initiate controlled deceleration if the car approaches the end of its normal travel range at excessive speed or outside expected deceleration profile. These switches operate within the supervisory system to ensure the car slows safely before reaching the extreme limits of travel.

Torque

The rotational force produced by the motor to drive the sheave and move the car.

Uncontrolled Car Movement (UCM)

A hazardous condition in which the lift car moves unintentionally without a valid command or independent of the intended control sequence.

UCM may occur due to brake failure, drive malfunction, mechanical imbalance, or supervisory failure.

Modern lift systems incorporate UCM protection devices that detect unintended movement and trigger immediate safety intervention, typically by reapplying the brakes or interrupting the safety circuit.

Up Collective

A control strategy in which upward calls are serviced during upward travel before reversing direction.

Variable Voltage Variable Frequency (VVVF)

A motor control method that regulates speed and torque by adjusting the voltage and frequency supplied to the motor.

INDEX

Made in the USA
Monee, IL
05 November 2026

67815698R00058